MW00823834

# Promise

# Promise

## A WIFE'S JOURNEY TO KEEP A PROMISE TO HER BELOVED HUSBAND

My life before and after my husband,
LCDR Ronald Lee Treinen died.
He was a Naval Intelligence Officer and he was stolen
from me by the military he proudly served and loved.

WRITTEN BY
# JOHANNA M. TREINEN

For more information, email Johanna.treinen.promise@gmail.com

ISBN: 979-8-9857693-1-9

# DEDICATION

Thanks to my husband, Lieutenant Commander (LCDR) Ronald Lee Treinen, for giving me his love, gratitude, attention and the best life a woman could ever want. He was a real officer and a gentleman. He loved God, his family, his country and patriotism. You, my love, will forever be missed and loved by me. The service has lost one of the best, most dedicated and loyal sailors there ever was.

I would like to dedicate this book to all the men and women who have served in the military, their spouses, children, mothers, fathers, sisters and brothers who have lost their loved ones due to the military's lies, betrayals and suspicious deaths. The careers of service men and women are often botched and derailed by false promises and unrealized expectations. These servants dedicate their lives at the expense of their own families and work very hard in order for all Americans to live in this country freely. People in other countries are dying to get the chance to set foot on this land, land of the free, for a better life. I thank you for your service. God loves you and so do I.

Sincerely and Blessings
Johanna Treinen

# ACKNOWLEDGMENT

First and foremost, I want to thank my Lord and Savior for being with me throughout this whole journey. He has carried me for all these years, one day at a time. He has shown His love to me, even when I was mad at Him and couldn't pray. He has given me "favor" when I was in the deepest valleys of my life and picked me up so many times that I couldn't even begin to count. He is the great I AM, my *rock*, my *comfort*, my *deliverer*, He is my God!!!

Thanks to Ken and Susan Treinen, Ronald's cousins, for their love, information, editing and support with my book.

I would like to thank my children, Rachel Sallee, John Gibson, Christina Messer, and Carrie Neely, for their support and input. But most of all, for giving their blessings in writing this book. Extra thanks to Carrie for your suggestions and taking the pictures for the cover. Thanks to Christina for helping me with my computer, reading, contributing, and editing my book several times and giving me hours of your time.

Thanks to my granddaughter, Caitlin Fitzhugh, for helping me edit my book and make things sound better. To her husband, Austin, thanks for setting up my computer with the business end of this book and getting me a new email address for my book.

(LCDR) Delta Whiskey for your listening to me and helping me with my questions.

(CDR) James Blasko, for asking questions about Ron's demise. Even when you got, "mind your own business," from the Navy.

Sally Sabow, for sharing her story about the murder of her late husband, Colonel James Sabow.

My friends at church, Elisabeth and David Viayra, Connie Briddell, Pastor Greg Bronson and his wife Elizabeth for giving their time in reading my manuscript and giving me their input.

Alejandro for the cover and formatting of my book. Putting up with my email messages, questions and my craziness.

Susan Michaud for proofreading, editing and answering my questions.

My friend and neighbor Karen Reno-Cobb for making herself available for extra editing and support.

Without the help of these people and others, which I may have forgotten to mention, this book would not have been published.

I AM TRULY BLESSED!

# INTRODUCTION

This story is about my life's journey before and after Ronald Treinen's death. He was my husband and a naval intelligence officer. He asked me to make him a promise: that if he ever died suddenly, I would have his death checked out. I did make him that promise and I'm sorry my love, that I waited almost 29 years, but now is the time for me to keep my promise. I will love you forever!

First, I share the story of my life with my abusive ex-husband, who was a marine. I discuss my stupidity on living with him for so long and how, through it all, I grew up and found self-love.

Second, although short, I found a wonderful life with my second husband LCDR Ronald Lee Treinen. He made me a happy woman and always treated me like his *queen*. We had six kids between us, he had three girls and I had one boy and two girls. Ron's naval career afforded our family the chance to travel and see much of our beautiful country.

However, my story is most importantly about how the Navy treated Ron. This man gave all his time, service and dedication to his career, all while moving his large blended family across the United States, seven times in nine years. Yet, he received lies, broken promises and an untimely death sentence. I never heard him complain…. Until he decided to get out of the Navy. Then, he explained to me why he was leaving.

Last, I write about how my family dealt with his death and faced the future without him.

# Table of Contents

# Part 1

## LOST LOVE

*When love is lost, do not bow your head in sadness; instead keep your head up high and gaze into heaven, for that is where your broken heart has been sent to heal.*

**Unknown**

My name is Johanna. I was the wife of a wonderful man, a Lieutenant Commander (LCDR) in the Navy. Now I am his widow. His name was Ronald Lee Treinen, and I became his widow on April 14th, 1993. In the 29 years since his death, I haven't been able to fulfill a promise I made to him. I can't provide answers or proof about his death. But the one thing I can do is tell his story.

1

Ron was born in Le Mars, Iowa on September 23rd, 1947. Ron's mother and father both served in the Army during World War II. His mother was a lieutenant in the nurse corps and his father was an enlisted man. After the war, his mother continued work as a nurse for the Le Mars public schools and his father worked as a short-haul trucker for a local trucking company.

Ron was the only child in his family. The last time Ron and I went to Iowa was to see his father who was in the hospital. A nurse came into my father-in-law's room to bring him lunch and recognized Ron. "Ronnie, is that you?" she asked. She stayed and told us stories about Ron, including this one.

Her family lived down the street. One day, Ron came down the road with his little wagon full of toys and told her mother that she could have all of his toys for her children if he could have one of her kids come home and live at his house. I guess there were a few kids at her house. Then Ron's dad told us about Ron being in the newspaper. It was Ron's fifth birthday and he went around knocking on some doors, asking people to come to his birthday party. He knocked on the Mayor's door. The Mayor's wife answered the door, she thought Ron was just the cutest thing she had ever seen and she called the newspaper, which published a little story about Ron knocking on doors.

## Big Business Booms In Le Mars' Younger Set Of Tycoons

Here's the story of a big business tycoon we heard this week that really puts a smile in the news.

Seems there is a young Le-Mars man named Ronny who will be celebrating his fifth birthday anniversary next week. He's the son of Mr. and Mrs. Peter Treinen of 221 First Ave. SE.

Monday noon Ronny got all dressed up, slicked down his dark hair and went calling on the neighbors, all unknown to his parents. He rapped on a number of doors until he came to one where the conversation went something like this:

"Would you please sign your name?" Ronny asked the friendly lady of the house, whereupon he whipped out a big tablet of paper and pencil.

When the neighbor inquired the reason for her signature, Ronny looked up with his large dark eyes and replied, "It's going to be my birthday next week."

Upon closer inspection of the tablet, the amused lady saw a dozen signatures, some written with the greeting "Happy Birthday to Ronny."

Still not quite getting the point, the neighbor put another question to our junior business man.

"Well you know next week is my birthday," he explained patiently, "and I thought they might like to buy me some toys and I'd like to have their names!"

Ron went to a Catholic school until he graduated from high school and went on to college in South Dakota. His goal was to become a lawyer, but the Vietnam War had started and Ron was going to be drafted. He told me that when he found out that his draft number was coming up, he joined the Navy before the draft board could make that decision for him.

In the early 1970s, Ron was stationed in the Philippines. While he was there, he met a young lady. One day she told him she was pregnant. He did love her and he asked her to marry him. Ron said he had set up a wedding date for them, invited friends, bought food and drinks and she never showed up for the occasion. He then set up a second wedding date and she missed that as well. According to Ron, she was gone for two weeks and he had no idea where she was. Then she just showed up as if nothing happened. She said she went to her family's house for a visit.

They eventually married and had a baby girl named Rachel. When ordered to a new duty station, Ron and his new family came to the States to San Diego. Ron and his wife had two more children: Sarah and Rebecca. As time passed, the couple found it harder and harder to get along. The marriage wasn't working. Ron learned that during the time he was away from home on two different deployments, his wife invited other men to live in their home. The kids would call them "uncle."

On the second deployment, Ron discovered that the affair his wife was having, was with a man assigned to the same ship, the USS Kittyhawk. I'm sure Ron must have felt so embarrassed. Before the ship departed for that deployment, his wife had left Ron and the girls to go see her lover off. Ron was on the deck, alone with his girls, waiting for her to return and pick them up to go home. The embarrassing part is that Ron was an officer and his wife's boyfriend was enlisted. Now, there is nothing wrong with being an enlisted man, but I think that Ron felt disrespected by a subordinate. Most people on the ship knew what was going on except for Ron, at the beginning.

Neighbors would tell Ron that when he went to work, she would bring a man into the house, and then he would leave before Ron got home. In the evening, while Ron stayed home and watched the kids, she would go out to be with her boyfriend. Ron finally got fed up and filed for a divorce.

*(Ron, on base with his girls. Rebecca in his arms, Sarah on the left and Rachel on the right.)*

My story begins in January 1982 when I was 29 years old. I was living with my three children in an apartment in Chula Vista, California. My first child was a boy named John. Then I had a girl, Christina. Finally, another girl named Carrie. I had been divorced from their father for three years. My ex didn't pay his child support. He was ordered to only pay $50 a month for three children. He was out spending his paychecks on beer, marijuana and other women. It seemed he would find a job that paid him under the table so he could lie to the court and tell them he had no job; therefore, he had no money. Then he told his friends and girlfriends that I was too lazy to get a job. Now, how was I going to find a job? I had no car, and with three kids, how could I pay for someone to watch them without any financial support? I was hurt and felt so helpless. For some reason, I cared what his friends thought of me.

I was sixteen years old when I met my first husband. A friend of mine had asked me to go with her to the club at the Marine Corps Recruit Depot (MCRD) base. There was a dance and she was meeting her boyfriend there. When we arrived at the barracks, she shouted for her boyfriend and another Marine came to the window. He was very flirty and invited himself to tag along. He was handsome and funny and gave me lots of attention. After several months, with reluctant permission from my mother, we were married. I was only seventeen.

At first, he was a fun guy, but three months into the marriage he became physically and verbally abusive to me. He beat me up at least nine times in nine years of marriage—always my fault, according to him. I didn't have the sense of a rock or the maturity to know that one time was more than enough. He broke my nose one time and my pinky finger on another occasion when I put my hands up to protect my face. When he broke my nose, I also ended up with two black eyes. When I was around eight months pregnant with my son, he got mad at me about something, grabbed me up off the bed and threw me down on the floor. Many times, he hit me because he said I would not shut up.

He was usually drunk, and I told myself he didn't know what he was doing. This behavior had been normalized. I grew up in the same type of atmosphere. My father was often intoxicated, angry and abusive. As I write this now, it makes me feel sick to my stomach. Readers, it is not okay for anyone to put their hands on you, especially when a man hits a woman. If you find yourself in this kind of predicament, please, run for your life. Get help for yourself. Only cowards beat women!

*(My ex-husband and I, on our wedding day)*

Of course, he was always sorry, and with tears in his eyes, he would say, "I'm sorry." He would then tell me it happened because of how he was raised. He too witnessed abuse between his parents. According to him, he and his father did not get along well. He had been a rebellious child and his father often used corporal punishment to correct his behavior. Many of us were disciplined in that fashion back then.

Early in our marriage, I remember visiting his parents one year on our anniversary. He told me he was going to see his friend and that he and his friend were going out to play a couple games of pool. He said he would come back after that, so we could celebrate our anniversary. I got ready for our celebration, and he never showed up. His mother said she couldn't believe he would do this to me. Mind you, she had warned me not to marry him because he had problems. But I wanted to be loved, and we got married.

When he got home around midnight, his mother was waiting to talk to him. She told him that I had gotten all dressed up for our anniversary and was now in my bedroom crying. I heard him come home but remained in the bedroom. He came into the room to go to bed. He was intoxicated. Again, he was sorry and asked me not to leave him; he needed me to get him to heaven. I said, "I can't get you there, you need to do it yourself by believing in Jesus Christ and asking him into your heart." His mother also told her son that love is like a tree, if you don't water and feed it, one day you'll look at the tree and find it has lost all its leaves, is withering away and soon will be dead. Smart woman, my mother-in-law. She always treated me as if I were one of her daughters, even after our divorce.

After that, I stayed with him because I wanted to be a good wife, and I thought that God wanted me to help him. But it got worse. He worked in Japan for about a year and left my son and me here in the States. He said we couldn't go with him, that this was an unaccompanied tour. But I found out later that my friend Gloria, who was also married to a Marine, went with her husband to Japan. My husband just wanted to go by himself so he could continue to cheat and I wouldn't know. He made me feel unloved and unwanted. When he came home, he found that I had lost sixty pounds. He was so excited. But then he got jealous. When we went to Tijuana to shop, some men stared at me even though I was dressed respectfully. I didn't even notice the men, but he came up behind me and whispered, "You could get any man you want until you take off your clothes." I didn't understand. I thought, *Do I gain my weight back?*

My stomach did look horrible. I had been five weeks overdue with my son, John. I was quite large. The stretch marks looked like outstretched lines of elastic all over my belly. When I finally started labor, it lasted for 50 hours. I was so tired and the pain was excruciating. I started to repeat aloud, "I wish I was dead." The nurse overheard me and talked to the doctor. They both came into the room and informed me that the doctor had ordered an x-ray to assess what was going on. After the doctor examined the x-rays, he returned and told me that he was going to perform a C-section. Because I had had polio as a child, it affected the ability of my cervix to open wide enough to give birth. My son was born at eleven pounds, eight and three-quarter ounces. He was twenty-three and a half inches long, so my stomach looked bad. I ended up putting up with my ex for six more years after that.

Even after we were divorced, he continued to put me down. My daughter Christina, who was about thirteen years old at the time, asked her Dad during a visit, "Why are you and mommy not together anymore?" He responded that he, "didn't like to (f@#%) fat women and your mother got fat." In the end, the only good things that my ex ever gave me were my three beautiful children.

*(Pictures of me when I was married to my ex-husband)*

13

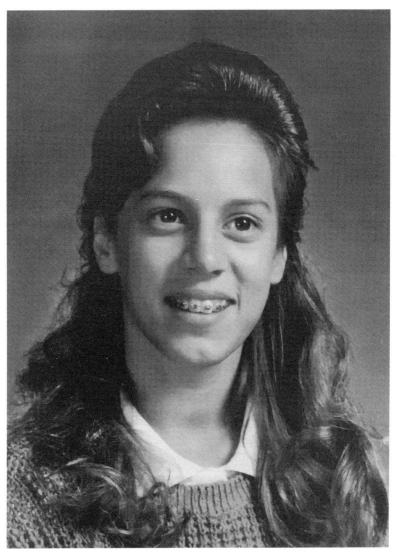

*(Christina at age thirteen.)*

After nine years of verbal, mental and physical abuse, I told my brother Cliff that I had just filed for a separation and was intending to get a divorce. My lawyer asked me why I wanted to file for a separation when I could just file for a divorce. That would save me paying twice for the same result, which made sense to me, so I changed the paperwork to a divorce. My brother asked me, "How are you going to live? Are you going on welfare?" I wasn't even thinking about welfare until he mentioned it. However, it was a great idea and I did go on welfare but not because I was lazy. I had no place to go, no money to feed my children and I was not going back to my parents' house. My dad was too strict, verbally and physically abusive. By the way, Cliff, thanks for giving me the idea. I survived with your help.

I will tell you that I was proud to have welfare as a resource to fall on. It helped me get on my feet. The welfare office sent me to school and I got my GED. I had only been seventeen years old when I married the first time. I hadn't finished high school. Then they sent me to train for a clerical position. I ended up getting a job with the Chula Vista Police Department. I worked first in records, then fingerprinting and then helped check pawn tickets for stolen goods. That job lasted for about a year. Eventually, the program that sponsored my job ran out of money. I got hired for another job as an accountant and then started selling diet drinks to earn extra cash. Welfare paid for sitters while I worked and gave me food stamps to feed my kids and myself. I am so grateful for all that I received from the government.

# Part 2

# FINDING LOVE AGAIN

*The greatest act of courage is not falling in love, but, despite everything, falling in love again.*

**Robin Wayne Bailey**

Oone evening in January 1982, I was relaxing and playing with my kids on the living room floor when the phone rang. I got up to answer it and it was Marie, my best girlfriend Belinda's mother. She had never called me before, so I was surprised when she asked me if I wanted to go out to the Oasis bar to go country dancing. I had just broken up with a guy over cheating and was feeling sorry for myself so I decided to go. I loved line dancing and two-stepping, especially when my dance partner could dance. It made me feel free and alive when my partner grabbed me tight and twirled me around the floor. I became energized and felt happy. I could feel the other person's rhythm and control of every move we made.

I called Elizabeth, my babysitter, and got ready to go. When Marie and I arrived at the Oasis, it was full of people dancing and having a good time. We found seats at the bar. Later, after we both finished dancing, I went to the restroom and came back and sat in my chair. Marie's chair was empty and after looking around the bar, I spotted her sitting across the room talking to someone she knew.

I sipped my drink and watched the people on the floor. A short time later the bartender came up to me and set a drink in front of me. Kahlua and milk, my favorite. "I didn't order this," I told the bartender.

"Yes, I know. The gentleman over there bought you this drink," she replied and pointed to a fellow who was sitting across the bar enjoying a beverage. I later discovered his name was Ron; he was out with his friend Jack. As I looked over at him, he held up his drink and nodded his head as if to say, "Hi." Then I returned the nod as if to say, "Thank you for the drink."

I've always avoided letting a guy buy me a drink because before the night was over he was thinking he was going home with me and that wasn't going to happen. I was not that type of girl. Not wanting to owe anyone anything, I returned the favor and bought him a drink, rum and Coke. He came over, thanked me and introduced himself.

"Hi, I'm Ron," he said. Then he asked me for a dance. I had so much fun dancing with Ron. He danced so well. He didn't step on my toes—not one time. I think we danced every dance that night. Did I mention he was very easy on the eyes? He was slender, had an athletic build and stood five-foot-eleven. He was clean-cut, with brown eyes, nice mustache and the sweetest smile. He was dressed like a preppy, in blue jeans with a pastel yellow buttoned-down shirt and penny loafers. Ron had taste. I was impressed. I thought he looked like a gentleman . . . maybe a businessman? Maybe a lawyer?

As it turned out, he was an officer in the Navy.

*(Picture of Ron around the time we met.) (Ron when he made Lieutenant Commander.)*

After the band stopped playing for the night, Ron asked me if I would go to breakfast with him, it was 2 am. I replied, "Sure, I'd like to, but I have friends with me."

"Bring them along," he said.

We went to Denny's and everyone sat around the table and talked about our lives—what made us happy, what made us sad, things we would like to do. And we talked about our kids. Ron had three kids like me, but all girls. Rachel was ten years old; Sarah was eight, and Rebecca was four. My kids, a boy and two girls, were almost the same ages: ten, six and four years old respectively. We talked about our divorces. Ron was in the middle of a divorce, and I had been divorced for three years.

Before we left the restaurant, I invited my friends and Ron to come over to my place that night and continue our conversation. We hung out for about an hour and a half until I grew tired and let everyone know I needed to retire for the night. I walked Ron to his car and kissed him goodnight, just a peck on the cheek. I think he was surprised that I didn't invite him to stay over. He said he had a really good time and asked if he could call me sometime. I gave him my phone number. He called the next day, Friday.

He asked me if I would like to go out dancing again. I said, "No thanks, I can't afford to go out too often and buy drinks, much less pay for a sitter."

"I didn't ask you that, I asked you if you wanted to go out dancing again. I'll take care of the rest," he replied. I was hesitant to accept because I didn't want Ron to know I was on welfare. I was still embarrassed about it even though I knew welfare was also a blessing and had done a lot of good for my kids and me.

Ron insisted. So we made a date for Saturday night, and we danced and talked until the early morning hours. It was like we had always been friends. He said that usually on Saturdays he would take his children to the beach or the park. *What a coincidence,* I thought to myself. *I do the same thing.* So we decided to plan an activity together with our kids so they could get to know one another. The next Saturday, we all met at the beach to enjoy time outside in the beautiful San Diego weather.

Ron and I were surprised to find out that the children already knew each other. They all rode the Joy Bus on Sunday mornings. The Joy Bus came through the local neighborhood and picked up children for Sunday school at a nearby church and then dropped them off back at home. Once we got together, he shared more of his stories with me. He wasn't a saint. He told me his wife had cheated on him, and he had also cheated on her.

After Ron and I had dated for about three months, I concluded that I had to break up with him. It was a good time to make a break. Ron was an instructor in the Navy, and he was leaving the next day to teach for two weeks in Hawaii.

I drove to his house and left a note on his garage door. I wrote that I was done seeing him and had decided to go back to my ex-boyfriend. "You should go back to your wife," I added. "You told me you still loved her three months ago, and obviously you still do, so go make things work with her. As for me, I'm done."

That night he went looking for me and ended up calling my mother, asking her for my phone number, which he couldn't find. He called me and asked me to tell him the real reason I was breaking up with him. I told him that my husband beat me up and cheated on me. To recover my self-respect, I had to learn to love myself. At this point, I was not going to waste my time with someone who was still in love with his ex-wife. I deserved to be treated like I'm somebody special—*like a queen*. I wasn't someone's plaything and I refused to be a doormat. He said he agreed; then he asked me if I would please put off breaking up with him until he returned from Hawaii. He was leaving the very next day.

While he was gone, he wrote to me regularly. I received about five letters consisting of at least 14 to 30 pages each. He wrote about his life and how he felt about me and that he would call me when he got back. He did call when he returned and he said, "Jo, I love you, but

I'm just afraid of getting hurt again." He told me that he didn't love his wife anymore. He loved me and that I did deserve to be treated like a queen. About a week later, he asked me to marry him.

I really loved him a lot, but I wasn't interested in raising six kids. So, we kept it friendly for a while just doing things on Saturdays and putting the kids on the bus on Sundays. It wasn't until some time had passed in our relationship that I found out that he didn't want to raise six kids either. Much later, we shared a laugh over this because raising six kids together was how it turned out.

Famous last words.

After a few months, we became more interested in each other and started to have a more intimate romantic relationship. We really fell in love. Ron Treinen was the best thing that ever happened to me. He was a handsome man, very polite and he did treat me like a queen. He would often buy me flowers. He'd take me out dancing and he always opened the door for me. On the rare occasion when he forgot, I would stand at the door and wait. He would give me a little grin and quickly remedy his mistake. When I would say "Thank you," he would always say "You're welcome" with a beautiful smile. When he was away on the ship or out of town teaching, we wrote each other letters. We ended each letter in a special way. I would write "Thank you," and he would reply, "You're welcome." He knew that when I wrote "Thank you," I meant that I appreciated him for loving me the way he did, for taking care of me and my children and for giving us the best life ever. I only have the letters he sent to me—not any from me saying, "Thank you."

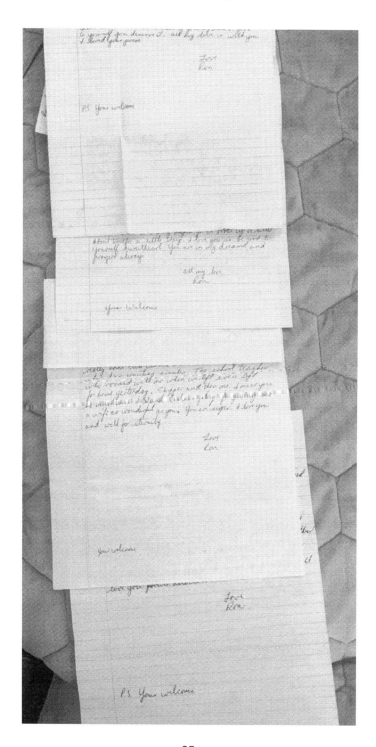

Writing letters is a practice that has more or less disappeared today. This is a time of computers and Zoom. However, after Ron died, I discovered the great advantage of writing letters. There is nothing better, than to read and relive, the words of love your loved one sent you in a letter, in their own hand.

Ron was always generous and made me feel special. He insisted on paying for everything, but there were some expenses I paid for because I don't like to take anyone for granted. One day, I took him to meet my parents. I asked my mother if I could borrow $50. I wanted to have some money because we were going somewhere and my payday was a couple of days away. When we left my parents' house, Ron said, "Jo, please don't embarrass me by asking your mom for money." He handed me a hundred-dollar bill. I told him I didn't want to take his money; I just needed a little money until Saturday. He didn't care. Ron made me take the money. I told him I would pay him back, but he declined. He said, "Jo, if I didn't want you to have it, I wouldn't have offered it."

# *Part 3*

## A DREAM MARRIAGE

*So it's not gonna be easy. It's going to be really hard; we're gonna have to work at this every day, but I want to do that because I want you. I want all of you, forever, every day. You and me....everyday*

**Nicolas Sparks**

Ron and I were together for eleven years—the best years of my life. We were married on October 23, 1982. All six kids were in the wedding. It was a beautiful sunny day; Ron and I said our vows at the North Island Naval Base chapel, in San Diego. We were so happy.

We honeymooned in nearby Las Vegas. Just after we arrived, I came down with the flu. I told Ron to go down to the casino and play cards or get some drinks. He was hesitant to leave me. I told him to go ahead because I was just going to sleep. He went down to the casino and returned in about an hour. He said it was no fun without me; then we both went to sleep. The next day we walked around town checking out the different casinos. After a couple of hours, the wind was starting to blow hard and sand swirled around us. It looked like fog and stung our eyes, so we decided to head back to our room. I was tired and wanted to take a shower and get a little nap before we went out to dinner. Ron headed down to play a little in the casino. When I woke up, my eyes were sore, tearing, and I couldn't keep them open. We went to the emergency room. The physician on duty said the sandstorm had blown sand in my eyes and under my contacts. He flushed my eyes with saline and gave me drops to make my eyes numb so they could heal. Not the ideal honeymoon we expected. But we were still in love.

*(Ron and me on our wedding day.)*

*(Left, top to bottom: Johanna, John, Christina and Carrie.*
*Right, top to bottom: Ron, Sarah, Rachel and Rebecca.)*

Our married life was also a military life. The Navy moved us around the United States seven times in nine years. One of Ron's assignments was in Washington, DC. He was a "detailer" who helped place officers in their next assignment, their next journey or their next ship. Ron shared the job with another officer. When their time as detailers ended, they would move to another duty station. Ron was supposed to go back to sea when he finished this job. The other officer was assigned to a joint college and then would move on to a joint job in Norfolk, Virginia.

A joint college unites different military branches, educating them on how to work together effectively in serving the U.S. government in defense matters. They learn how to plan and operate together during wartime or in specific combat situations or national security events. They are a large part of our national and global service, incorporating governmental and non-governmental bodies, as well as international organizations.

The other detailer's wife was pregnant and ready to give birth. He was torn about what he should do. He told the command, "My wife will be giving birth anytime and she needs me to be there with her." Ron was asked by a commanding officer if he would be willing to replace him and take the school posting. As an incentive, Ron was promised that after the school was over, he could pick the ship he wanted to be on and become the executive officer (XO). When Ron came home that evening, he told me about the new deal. He said that if he accepted, he would choose the ship USS Fort Fisher because the duty station was in San Diego, near our home. Someone had to fill that billet for the other man, and it might as well be him. Given the enticing offer, Ron then got things in order and let command know that he agreed to take the school.

Ron went to school in Quantico, Virginia. It is about 40 minutes outside of Washington, DC. Quantico is the home of the largest U.S. Marine Corps base, Marine Corps Base Quantico. This base is also the site for other agencies including the FBI Academy and the Naval Criminal Investigative Service (NCIS).

After he finished school, Ron's whole life changed. None of what he was promised in DC happened. Instead, Ron was told that because he went to the school in Quantico, he had to do the joint job. He was assigned to work as a naval intelligence officer in the War Room in Norfolk, Virginia.

The officer who had promised Ron his choice of posting when Ron was in DC said that he tried to get naval commanding officers to honor what Ron had been promised. The appeal didn't go anywhere. Slowly, we came to understand that the door was closed on honoring any deal Ron had accepted when he was in DC.

Ron and I were deeply disappointed. It was hard for both of us to reconcile that the same Navy that Ron had honored for so many years and had served with so much pride had reneged on its commitment. We felt deceived and betrayed, but Ron made the best of it. After Ron and I talked about what happened, he never talked about it again nor expressed any feelings.

We were on the move again—this time to Norfolk, Virginia. We rented a house in Virginia Beach and found another school for the kids. Three of the kids were teenagers by this time. With each move, it became harder for them to adjust to new communities, cultures, customs, neighbors, routines, friends and schools. It was also hard on the younger kids. As an incentive, I gave them a dollar for each new friend they made. Thank goodness that worked. Most military men and women and their dependents know what I'm talking about.

Ron started his new job. Shortly after, he and I were supposed to go to a "Hail and Farewell" event for him and others leaving or arriving at the base. On the day of the event, Ron came home to prepare to go out. I was lying in bed. I wasn't ready to go. He inquired why I was lying down when we had a party to go to. I told him my clothes were getting tight and I didn't want to go. I had gained weight from all the stress in our lives and I was afraid that everyone at the event was going to look at me and say things like, "I didn't picture him with her." Although he was ready to go and had his shoes on, Ron looked at me and said that he wasn't going either. He got into bed next to me and held me so close. He said, "If you don't go with me then I'm not going either."

I said, "But you have to go, this Hail and Farewell is for you."

Ron replied, "I know, but you are a part of my life and if you don't go, I won't either! We do life together." I got up and got dressed and we went together. I wanted to support such a handsome and kind man who always made me feel loved.

After two years on the job, Ron was once again up for promotion. He always made it the first time. However, this time was different. Ron was passed over despite having met every requirement and the wonderful reports written about him. When he told me about it, I saw a couple of tears come from his eyes and I felt so bad. I told myself that if Ron had received the job he had originally been promised, he would've made it.

After that, Ron wouldn't talk much about it except to say that this was the last time that he would bend over for these guys. "I am going to get out of the Navy," he said. He had served for 22 years and leaving was a hard decision for both of us. I didn't want him to give up. And there was a financial reality: Ron was the provider for eight people. Changing jobs was a big financial risk.

In spite of all the uncertainties and risks, Ron still chose to leave. I told him that I would support him if leaving was what he needed to do. This is life in the military as it is life in every employment sector: corporate, public or private. It happens to a lot of people. There are so many people upset about the way their jobs or careers turned out. They try to hold on for fear of wrecking their life or the lives of those they care about. They keep their mouths shut. But as you can see, it gets messed up anyway. The whole family suffers. I pray for all of you who have had to endure this, especially those who are in the Armed Forces.

A few months before Ron got out of the Navy, I walked into the kitchen. He was sitting at the breakfast table reading his Bible. It was quiet and I didn't want to disturb him. I started to fix breakfast. All of a sudden, Ron said, "Jo, will you promise me something?"

"Sure," I replied.

"If I ever die suddenly, would you please have it checked out?"

"Why," I asked.

"Just promise me, please."

I promised and left him alone. But his request and my promise have stayed with me ever since that day.

Ron never talked to me about his job. But I knew how loyal and dedicated he was to any job he ever had and to the Navy. I remember when we first got together, I put on his Navy hat and he got upset and asked me to take it off. I asked him why and he said that I wasn't showing respect for the uniform. To Ron, the USA was the best country to live in and even though things weren't perfect, he would always fight for his country. He was so patriotic. He took us all out for Hands Across America so we could see and participate with other Americans, standing hand-in-hand, showing our children what it looks and feels like to be American. He took the family to downtown Washington DC. We visited a number of the museums and monuments that showcase the history of the country. He took us to NASA to view the space program. Ron wanted the kids to understand this country's development and achievements. This was the Ron I knew, the Ron who loved his country and who served the USA with pride and dedication.

Ron's retirement from the Navy was coming up in June and a retirement ceremony was planned for him. One day when Ron and I were talking about his retirement, he said he was going to say something about hating his job. I never gave it much thought because he was the one whose service was being recognized and I felt he could say whatever he wanted—just so he could leave with a little dignity.

I didn't go to his Retirement Farewell because I was overwhelmed with moving the family back to San Diego where one of our daughters was getting married. I had to prepare for the wedding. I loaded the kids in the van and drove across the country, home to San Diego.

Ron gave a speech at his retirement ceremony. The speech was taped and later I was able to view the tape. He said a lot of nice things about me and my handling of the household and ourmoves and how I made things easier for him. Then he said to the other people in the room that he hated every day he walked into this job. He proceeded to tell the men that when life gets you down and you feel you can't take it anymore, take it to God. He said, "That's how I made it through and God will help you do the same." I never knew he felt that badly about his job and he never let on that anything was that wrong. It startled me because my husband was always so loyal to the job and the Navy. He didn't deserve what happened to him.

# *Part 4*

# END OF A DREAM

*Though lovers be lost, love shall not; And death shall have no dominion.*

**Dylan Thomas**

After the long cross-country journey with the kids, I finally made it back to San Diego, California. San Diego was so warm, and I thought how nice it was that Ron and I would now have our forever home in this pretty location.

Our daughter Rachel's wedding was beautiful. She married Greg, the boy who lived across the street from our house in San Diego. At the time, she was in college at San Diego State University and he was working as an accountant. Ron and I couldn't have asked for a better partner for Rachel. Greg had a strong faith in God, a great work ethic and a kind, playful personality. We were so proud of them as a couple. They are still married and have three beautiful children.

*(Ron and Rachel on her wedding day.)*

*(Left to right: Ron, Johanna, Rachel and Greg.)*

*(Back, left to right: Ron, Rebecca, John and Sarah. Front, Left to right: Johanna, Rachel, Christina and Carrie.)*

After the wedding, the family moved back into the house we owned. Ron and I had rented it out during the time we were in Virginia Beach. The renters still had time on their lease, so we lived in my parents' house until after the renters left and the house was cleaned. The kids were back in school and Ron had signed up for law school. Ron had always wanted to be a lawyer. He was often studying and falling asleep on the couch. I would be watching television, and he would be fast asleep beside me, books in hand. I would wake him up and tell him to go upstairs to bed. He would say, "No! I am studying. I just wanted to keep you company."

This episode brings back memories of times when Ron was still in the Navy and wanted to "keep me company." One time Ron had dozed off and I had my red nail polish out. While he was sleeping, I painted his toenails fire engine red. He woke up later than he had planned, quickly got dressed and walked out the door to get to the ship on time. He said he saw the polish but didn't have time to take it off. When he got to his stateroom, some men were visiting the ship and Ron had to share his room with one of them. While he and the visitor were both in the room, Ron changed from his street clothes to his work clothes. As he was changing his shoes, he removed his penny loafers and the visitor saw the nail polish on Ron's toes. Ron said the guy looked at him strangely. Ron looked up and simply said, "My wife!" The last time Ron kept me company and fell asleep, he woke up with bandages stuck all over his body. He never felt me putting them on him. The next morning, he just laughed and said "I love you. You make me laugh!"

Ron found a Bible study group at law school and made good friends. They enjoyed discussions together about both law and the Bible. He exercised every day. He was a runner. When he went to the college to attend class, he would also go to the Naval Training Center (NTC) and run on the field, usually late mornings or around lunchtime. He said that running relaxed him.

On our last night together, Ron was sitting on the couch, reading the newspaper and I was sitting on the loveseat making fabric apples for one of my craft projects. It was a Wednesday night and the kids were leaving to attend a youth group meeting at church. After the kids left, Ron tried to get my attention. "Hey!" he said to me, lifting his eyebrows up and down. This was a signal to me that, with the kids gone, we could go upstairs and make love. When the kids were around, we used the, "the dresser needs fixing," signal. Ron would say, "The dresser needs fixing, come and help me with that." It was our cute way to communicate our need for one another. The reason I mention this was because Ron always made me feel wanted and needed.

That evening, after "fixing the dresser," Ron looked into my eyes and said, "Jo, I wish you knew how much I love you. You mean everything to me." About a half-hour later, we went back downstairs. Ron went back to reading the newspaper and I worked on my crafts.

"Ron, you're either an angel or an alien," I said looking at him.

"Really? Why?" he said, smiling as always.

"Because every time you make love to me, you make me feel like it is a pleasure."

"It is."

When the kids got home, they did their homework and got ready for bed. We all turned in for the night. During these moments with Ron, I often realized how much my life had changed for the better over the years. With my ex-husband, there was constant abuse and disparagement. Ron never called me names, never hit me and always had wonderful things to say to me. Every day, Ron would greet me with, "Good morning beautiful." Oh, how I miss hearing him say that!

On the day Ron died, I got up in the morning to kiss him goodbye before he left for school and he gave me a big hug and kiss. Then he went out the door. For some reason, I felt I needed to get another kiss from him but I was still in my pajamas and didn't want the neighbors to see me. I stood by our front gate to see if he was still there but he was already moving the car into the street. "Oh well," I thought, "I'll just give him a kiss when he gets home."

I started my day by going upstairs, taking a shower and getting dressed. I then headed out to the swap meet to look for curtain material, so I could make new ones for the house. I bought some fabric and headed home. I was unrolling the material to make sure I had bought enough when the phone rang. It was about 5 p.m. The female voice on the other end asked me if I was Mrs. Treinen. I said, "Yes," and she asked me to hold on the line for a minute.

Then she came back and said, "This is Sharp Cabrillo Hospital. Are you Mrs. Ronald Treinen?"

"Yes," I said again and continued with, "Are my girls all right?" My first thought was that one of my three daughters was injured in a fall while cheerleading. The nurse didn't respond and put me on hold again.

When she resumed talking she asked me, "Is your husband's name Ronald Lee Treinen?" Frustrated, I answered, "YES, IS HE DEAD?"

"Yes," she replied and put me on hold again. When she came back on the line she asked me if I would come to the hospital.

I hung up the phone and let out a blood-curdling scream, "Oh God! Why?" I dropped to my knees. My first thought was that this must be what my husband meant when he asked me to check into his death. I felt as though my whole world had fallen from under me. My life was over. I couldn't fathom how I was going to make it. The pain was so intense, I wished someone would knock me out so I wouldn't have to deal with this heartbreak. Thinking about everything drove me nuts. But I have children who just lost a father and I had to be the example. I had to pull myself together. I asked God to please help me with all of this chaos.

I called the school and told the kids to come home. Then I ran outside. I didn't know what else to do. I was home alone and running around like a chicken with its head cut off. Finally, I ran down the street to my friend Janie's house. Her husband was also in the military. She opened up the door, and I cried, "Ron's dead!"

"Who?"

"Ron!"

Shocked and not knowing what to do or say, she just hugged me with dismay and said, "No! It can't be!" After a bit, I ran back to the house because the girls would soon be arriving home from school, and I had to be there. When I heard the car with the girls approaching the house, I went out to meet them; Christina, Rebecca and Carrie.

Christina was the first to reach me. I dropped down in the middle of the yard and wept, saying, "Daddy is dead." We hugged each other and cried.

At that moment, I understood that the dream Ron and I had together was really gone. It was as if the best dream I ever had was a balloon and someone had just popped it—boooommm. The dream had seemed so real and then I woke up. My dream was over. There was never going to be an "Our," an "Us," or a "We" anymore. I was alone. Somewhere inside of me, I was pleading, "Give him back, Lord! I can't do this without my Ron!"

Just then, my brother Mark drove up to my house, hugged me and prayed for me. I calmed down enough to get us all together to go to the hospital. We went to my mom's house on the way to the hospital. My brother Clifford lived next door to my parents. He said he would drive me to the hospital because I was not in any condition to drive.

As I got to the hospital I felt as though God had wrapped my heart in cotton. I felt supernaturally comforted. I went into the room where Ron's body was lying. When I touched him, he still felt warm. I was confused. I was told that he had died in the morning, and it was now about 6 p.m. I didn't understand how his body could still be warm. There was a tube coming out of his mouth, which I guessed was placed there for breathing, but the tube wasn't attached to any oxygen.

"When Elisha arrived, the child was indeed dead, lying there on the prophet's bed. He went in alone and shut the door behind him and prayed to the Lord. Then he laid down on the child's body placing his mouth on the child's mouth, his eyes on the child's eyes and his hands on the child's hands. And as soon as he stretched out on him, the child's body began to grow warm again! Elisha got up, walked back and forth across the room once, and then stretched himself out again on the child. This time the boy sneezed seven times and opened his eyes!" 2 Kings 4:32-35 (NLT)

As odd as it sounds, I wanted to be alone with Ron, doors locked, windows covered. I wanted to place my body on Ron's body and pray to God to bring him back, just like Elijah had done in the Bible.

I didn't let myself do this. I worried that if the hospital staff saw me, they might try to sedate me or place me in a hospital or take my kids away. But to this day I regret my fears and my decision. I have come to believe that there is often a need in humans for one last physical contact with a person they love before their loved one becomes a memory. It's not unnatural and doesn't mean the person isn't of sound mind. Perhaps it's just the opposite.

"For God hath not given us the spirit of fear; but of power, and of love, and of a sound mind." 2 Timothy 1:7 (King James Version)

After the kids had visited Ron's body in the room, we started to leave for home. Before we left, the hospital nurse handed me the gym clothes and shoes that Ron had on when he died. For me, this answered the question of where Ron was when he died. He was doing his daily routine at the NTC.

My son, John, told me that on the way home, he could see me from his car. John said he felt so sad for me because he saw that as I was driving I was clutching Ron's tennis shoes to my chest and just bawling—something I don't remember, but he swears by it.

The night before the wake, Sarah said that her mother, Ron's ex-wife, wanted to help me select things for the funeral. I agreed since it was a request from Ron's biological daughter. The ex-wife came with me and the girls to make the funeral arrangements. I felt that if Ron's girls participated in some way it would help them grieve. The girls and I created several collages with photos of Ron's life with the kids and me and put them up throughout the funeral home. It looked beautiful. I would come to find out later that his ex-wife just wanted attention and recognition as the first wife.

At the wake, there were a lot of people, both friends and family. Ron's ex wife arrived while I was out in the lobby talking to his best friend, Jack. She and her daughter, Rachel, went around and looked at the photo collages. Rachel was proud to share the photos of her father. Later, Rachel told me that her mother was very upset because she wasn't in any of the photos. As the eldest daughter, her mother blamed her for not making sure that she had been represented, in pictures, as the first wife. I wish I had been there with Rachel when her mother said this. I would have stepped in and reminded her that it was our place to comfort our children at the time of their father's death, not make it about ourselves. Ron's ex-wife never showed up for the funeral the next day.

During the wake, when I stroked Ron's hair, I could feel all the staples in his head from the autopsy. It was a hard discovery and I quickly pulled my hand away. I had wanted to check his body out to see what they did for his autopsy. I was never consulted on the decision to conduct an autopsy. I am not even sure who made the decision. A hospital official? A naval base official? When I got the autopsy report it said that the cause of death was a blockage in his heart. A couple of other things in the report caught my attention. The report also said that Ronald had pink lungs. How could that be? He had smoked since he was a teenager. I asked a doctor friend of mine, Dr. Zhao, if someone who had been a smoker most his of life could have pink lungs? Dr. Zhao said that would have been an impossibility. The report also stated that Ron had a needle mark on his neck. Why was that there?

We all have to go through loved ones dying and we all grieve differently. Without God in my life, I would be six feet under. When the funeral was over, we all went over to my house and celebrated Ron's life, as Christians do. Christians believe that after humans die on earth, their human body deteriorates and becomes part of the earth. But their spirit survives and the spirit of those who are saved by the Lord resides for eternity with Him in Heaven. We also believe that in Heaven our spirit will be united with the spirits of our loved ones. Ron and I will again reside together forever to celebrate God, Jesus and the Holy Spirit!

"We are confident, yes, well pleased rather to be absent from the body and to be present with the Lord." 2 Corinthians 5:8 (New King James Version)

Even though there were kind thoughts and cards from many friends and family, seeking to comfort me, I went through the day of the funeral in a fog. I knew that God must be watching over me because, In the midst of what was for me the greatest of life's tragedies, He kept me calm I got through the day, doing what I could to make everyone comfortable and see that we had plenty of food. My friend Cindy took me outside to talk and asked if I was okay. She said that she was surprised at how well I seemed to be dealing with everything. I told her I was dealing with everything because I wasn't feeling much.

I also wasn't eating much. Later, when I admitted to my brother Cliff that I hadn't eaten in three or four days, he told me if I didn't eat something, he wouldn't come over to the house ever again. I took a few bites. Even though almost every memory of that day is a blur, I do remember that during the funeral's twenty-one-gun salute, I would jump in my chair at each shot, even though I knew the shots were going to happen. Each shot of the rifle felt like they were actually going through me. My mind kept thinking about Ron, and curiously, I was thinking that he didn't look like himself in the casket. The body looked like his body, but his face looked different. Others had commented to me that the body did look like Ron. I heard them, but I was still bothered by what I saw.

Once the celebration of Ron's life was over and friends and family started to leave, I cleaned up and headed to bed. I found it so hard to lie in bed without my partner; I would hug his pillow and cry. Sometimes I would take some of his clothes with me and try to smell him. After the kids returned to school, I was alone more and more. Five o'clock in the evening was the worst. This was the time Ron usually walked through the door, coming home from school. But Ron wasn't going to be walking through the front door that day or any other day. His dog, Cord, a toy poodle that Ron hadn't initially wanted but eventually fell in love with, would sit at the front door, waiting for his master to come home. This lasted for months. Who would have guessed that animals grieved as humans do?

*(Ron and his dog Cord.)*

I started to break down. I was constantly sleeping and crying. My way of coping was to sleep. Alcohol was not the answer for me. If I did have a drink, it was coffee liqueur with milk, about five glasses a year. I didn't take drugs. The next best thing was sleeping. I would have to stop myself from crying because I would cry so much that I felt drained of my energy and couldn't breathe. I would stop and then I'd start again.

I became a recluse. I had no life, no husband, no dreams. I had wonderful children, but I was still alone. I asked the Lord why not take me? At night, in my room alone, I would open up the window and yell, "Ron, where are you?" as I brushed the tears from my face. I know the neighbors must've thought that the crazy lady was at it again, screaming out the window. I would cry so much that I felt as if I had run a marathon. For hours, I would lie on the bed and sleep while hugging Ron's pillow. I would wake just before the kids got home from school and when they arrived home, I'd pretend everything was all right. But, no, it wasn't all right, and it would never be all right again. To this day, the thought that my life will never be the same again still hurts.

One morning after Christina left for school with Carrie and Rebecca, I went upstairs to my room, knelt down and started to pray. My tears started flowing like a river, I was screaming, "God, just let me hug Ron one more time!" I knew in my mind that God wouldn't permit that to happen. He knew I wouldn't be able to let Ron go. Suddenly I heard footsteps on the stairs; I tried to stop crying but couldn't. I opened the door and Christina was standing in the doorway. She had heard me crying and asked if I was all right. I replied that I was all right, but I just wanted to hold Ron one more time. Christina hugged me and I cried for a minute or two.

Then I took a deep breath and thanked her for her concern and for hugging me. I told her that I was okay and that she and the girls should get to school. I didn't want them to be late. After Christina left, I was so grateful that she and all my kids are caring and have such good hearts.

Thoughts plagued my mind day in and day out. I was a widow. I was grieving. My world would never be the same. It could have destroyed me, but my faith and God's love helped me keep it together. Did you know that God catches every teardrop and keeps it in a bottle? I'm sure God has filled a lake with my tears by now.

"You keep track of all my sorrows. You have collected all my tears in your bottle. You have recorded each one in your book." Psalm 56:8 (New Living Translation)

Sam, a friend of my brother Mark, heard about Ron's death from his neighbor, Marcos. Sam confirmed with Mark that it was my husband who had died. Then Sam passed along an eyewitness account from Marcos. Marcos had been at NTC that day, working at his job in the gym when he noticed people gathering outside the building. When he went out to see what was happening, he saw Ron lying on the cement. Marcos started CPR on Ron and told someone to call the ambulance on the base. The person returned and said there were no ambulances on the base right then. Someone then called MCRD, the Marine Corps Recruit Depot—no ambulances there either. All ambulances were elsewhere.

While waiting for an ambulance, Marcos continued giving Ron CPR. Marcos said that during that time Ron took a couple of breaths on his own. An ambulance finally arrived 40 minutes later. According to Marcos, the ambulance attendants and EMTs didn't seem in a rush. They talked among themselves, put Ron into the van and drove off to the hospital. The lack of ambulances is still puzzling. The two bases, NCT and MRCD, are next door to each other. How could all the ambulances be gone at the same time? I later learned there was also a civilian ambulance and fire department right outside the gate of NTC, but the fire department was never contacted. No ambulances. Is it a coincidence? I don't think so!

Sam gave me Marcos' phone number. I called him to schedule a meeting, so I could hear his account for myself. We met at a Black Angus restaurant and Marcos allowed me to tape the meeting. I agreed not to use his name since he was still in the military. Marcos told me that Ron did not have his military ID card on him when he was found, nor was it in his gym locker.

Apparently, there was a rule that required those who run or workout on the base to have a military ID on their person at all times. Ron's military ID was never located. I am confident that Ron would have had his ID on him. Where did it go?

Death was so hard for me to deal with. Once the clouds of cotton were removed from my heart, the pain of losing Ron intensified. I became angrier and angrier at God! I remember going to visit my brother Mark's church to hear a special speaker at the Sunday service. I don't remember who the speaker was, but during the service, the church honored a couple for being married 50 years. It upset me. Talking to myself, I said to God, "Yeah you give them 50 years, and I only got 11."

I was so angry that I stopped praying for a while. I couldn't understand why this was happening to me. I had finally found a guy who loved me with all his heart as I loved him with all mine. *I was happy, God! Where were You when I needed You? You could have saved him.* Then I listened for God to answer, hoping I would get an answer. There was nothing, dead silence. A woman at my church tried to help. She kept writing to me, telling me how much God loves and cares for me. She would send me beautiful poems and cards about God's teachings. One card, in particular, got my attention. I had wanted God to answer me, and I found an answer in this card. The card said: "God, where were You when my husband died?" And God replies, "IN THE SAME PLACE I WAS WHEN MY SON DIED." WOW! God knew I wasn't letting Him speak to my heart, so He sent this woman to me as His messenger.

I decided to see my psychologist, Dr. Jacobs. I needed his help to ease the pain. I initially saw him when I was around twenty-seven years old and wanted to fix myself. I needed to know why I always picked abusive partners.

I was able to talk to Dr. Jacobs and share my feelings with him—what I had done and what I was afraid to do. He was compassionate and caring and helped me learn to love myself. It was his counsel that gave me the ability to file for divorce from my first husband.

Now he was helping me cope with losing Ron. I told Dr. Jacobs that after Ron died, the pain of his loss was so severe that I sometimes had thoughts about running my car into a cement wall as I drove down the freeway. I would never actually do that. I was trying to convey to him how much pain I was in. However, sometime later, I received some insurance papers in the mail that stated I was suicidal. I was furious. I went to see Dr. Jacobs one more time and asked him why he had said that about me. I told him that I trusted him by telling him my feelings and asked him why he was telling the insurance people that I wanted to commit suicide. He explained that it was because the insurance would pay for more visits, so I could continue to see him and get more help. He also said that it was his duty to let people know if a patient may commit suicide.

Seeing this statement on paper was something I never wanted my kids to see. Even though I had sometimes felt that way, it always passed. I would never do that. I gathered my kids together and said to them that if someone ever told them that I died from suicide, it was a lie.

I would never leave that memory for my children. I love them too much to give them that memory of me or to give them any thoughts that suicide is an option when things get rough.

I stopped seeing Dr. Jacobs. Later I did send my married children to him because he is very good with relationship problems and teaching love of self.

# *Part 5*

## STRANGE HAPPENINGS

*"Jo, will you promise me something? If I ever die suddenly, would you please have it checked out? Just promise me, please."*

**LCDR Ronald L. Treinen**

From the day Ron died, I have suspected that he didn't die from natural causes. Immediately after learning of Ron's death, I remembered his request to check into his death if it was sudden. Ron was worried. The big question, both then and now, is "Why was he worried?"

Odd circumstances surrounding Ron's death started to pile up almost immediately. His missing military ID, the 40-minute delay in getting an ambulance to the scene, the hospital's hours-long delay in informing me of his death, the autopsy the hospital performed without my permission and Ron's changed appearance in the casket. And there were more to come.

At one point, it seemed like my phones were being tapped. I would get phone calls and when I answered, there was no caller on the phone, only soft country music playing.

A couple of months after the funeral, I left the house and ran some errands while the kids were in school. I returned home and went up to my bedroom. On my bedroom floor, I found three award certificates that Ron had received from his last job. The documents were out of their frames, nine pieces lying on the floor. I knew that one of the places those awards were originally stored was on the shelf on Ron's side of the closet. Has someone broken into the house? I stopped, looked around and didn't see anybody, but I knew someone had been in my house. I also knew that I was meant to see the awards. How anyone would be able to get in the house is a mystery. I had changed all the locks to my house right after Ron died.

What was I intended to see? Who wanted me to see it?

A year or so after Ron's death, my youngest daughter, Carrie, and I were driving on the freeway and a white SUV was to my right in the lane next to my car. I've had this type of experience before. When I would drive faster, the other car would speed up. Or when I would go slower or change lanes, the car next to me would do the same. Many times, when this has happened to me, it would be a friend or family member who wanted to say hello.

I finally looked over to the right and in the SUV was Ronald! The spitting image of him! He smiled at me too! The smile I knew and loved! I was shocked. Thank God, my daughter was with me and witnessed the same thing I did. I asked Carrie to look over at the car on the right and tell me who she saw. She gasped and said, "Mom! That's Dad!" I tried to stay with him. But it seemed that once he was sure we had seen him, he exited the freeway. It was rush hour and too many cars were in the way. I couldn't follow him.

Later I was at my daughter Rachel's house with my other girls and some of their friends. I shared with everyone what Carrie and I had experienced on the freeway. One of the friends asked one of my girls, "Do you think that your dad would get another identity to save all of you?" They all answered, yes, Ron would do that. Yes, most definitely.

*(Me and Carrie)*

Too many things were happening. Were they just coincidences? Was I just a grieving wife missing my husband? No! Ron wasn't a paranoid, fearful man. But something about his demeanor, on that day, when he made me promise, was direct and filled with importance. My children however were terrified for me. They didn't want me to say anything to anyone. If the military killed Dad, what would stop them coming for you, they would say. To make them feel safer, I laid off looking into his death for many years and it has cost me.

Twenty years later, I reconnected with another Navy wife who I had known long ago. Her husband, LCDR Delta Whiskey (Navy Seal) was in the garage and I went in to say hello. I started sharing some of my stories about Ron with him. At one point, I asked Delta Whiskey why these things had happened to me? He replied, that the military wanted to scare me. Maybe my children were right?

Once, my daughter Christina and I went to a three-day conference featuring Joyce Meyer. In the hotel where we were staying, a group of military men had gathered, attending a conference. I can't remember how we started talking about Ron, but one of the men told me that sometimes people are given different identities in the military and sometimes they will make up somebody's face to look like another's. That surprised me. I hadn't mentioned to the man that Ron did not look the same at his funeral. The man also said that there are military experts who can make a corpse look like somebody else and they can do that really well. For many years, this has given me a small hope that Ron is alive and living under a different Identity.

I also told the soldier that my husband was on base, running, when he collapsed in front of the gym. The soldier told me that the military can easily get rid of anyone they wanted. They could have sent someone to run alongside Ron, just another soldier, nothing suspicious. He could have run by and hit Ron with a needle or a ring or something. It would appear as if he had a heart attack and the autopsy report would say that he died from a heart attack. The military does this kind of thing.

For the sake of my children and their families, I had not done anything that could put my life at risk. But, I had promised Ron. It has been 29 years since Ron died and I have felt miserable because I had not yet fulfilled my promise. That decision, however wise it was then, did cost me. I became less social, mostly having relationships with only close family. I had many health issues over the years, which could have had psychosomatic causes.

A short time ago, I decided that there was an action I could still take. I could tell Ron's story. My inspiration was a friend named Sarah Sally Sabow. Her story is similar to mine. She is still trying to find answers as to why and how her husband, Colonel James Sabow – USMC, died. Sally had gone to church and when she returned to her house her husband wasn't watching TV as she expected. She found the TV turned on but her husband was missing. She looked around the house and he wasn't there. She then went to a window and looked outside. In her backyard, Sally saw her husband; he was lying on the ground and was unresponsive. After the investigation the military, told her that her husband had died by suicide.

When I talked with Sally, she said there was no way her husband would ever commit suicide. He wasn't that kind of man.

She told me not to waste my money trying to get answers about my husband. Sally and her family spent hundreds of thousands of dollars on lawyers and expert resources and on exhuming her husband's body. She still doesn't have the acknowledgment she seeks, that her husband's death wasn't by suicide. She believes he was killed. There was one more avenue Sally and Colonel Sabow's friends and family decided they could take. They could tell James Sabow's story. Sally's brother-in-law, who is a doctor, and three other friends took on the quest by writing a book describing what happened to Colonel Sabow and their efforts to find the truth about his death. The book is called Treachery, Murder, Cocaine and the Lucifer Directive by Robert O'Dowd with Brian Burnett, Robert McLaughlin MD, and John David Sabow M.D.

Sally and I have become friends since our first conversation. I told her that I was also writing an account of what happened to Ron. In this way, I could, to some extent, fulfill the promise I made to Ron. At least I could tell his story. I also wrote this story to encourage others in the military who have had similar experiences to speak up and tell their stories.

I have looked into Ron's sudden death and considered all the facts, testimonies, and information I had access to. I also considered my own experiences described earlier in this story. Based on the knowledge I have, I had to conclude that Ron very likely did not die of natural causes. Instead, he was deliberately killed. Like Colonel Sabow, Ron's work during his last tour of duty in Naval Intelligence and in the War Room somehow influenced his death. In some way, Ron became a threat to the military. He knew too much. He wasn't complacent about what he knew. Whatever he had learned bothered him. I know Ron said he hated every day he stepped into that job. He disclosed his dislike and talked about it at a public reception attended by Naval Commanders. I have the tape of that reception and I saw that when they panned the camera over to one of the admirals, his face showed that he was very upset by Ron's remarks.

When he asked me to check out his death, it was around the same time Sally Sabow's husband was killed. If I had to guess, I would say it just might be connected to Sally's story. Her husband's death had something to do with drugs and Ron's job in the War Room had to do with drugs, amongst other things. So I thought possibly the two deaths were connected.

The fact that Ron ended up in a job he hated is one of life's greatest tragedies. He tried to help out a fellow officer by taking that officer's school assignment. In return for accommodating this officer, Ron's commanding officer made promises he couldn't keep. The only real winner in the deal was the fellow officer Ron shared a job with. LCDR Ronald Treinen, who loved his country, was faithful to his family, faithful to his service in the military and a wonderful man of God, died. Why this happened will always haunt me.

So that is the story of Ronald Lee Treinen. I hope that this story will encourage others in the military who have had similar experiences to speak up and tell their stories. By sharing our stories, perhaps we can change the outcome for others in the same circumstances. We need to get our governors, assemblymen and congressmen to notice the wrongs that are happening to our men and women in the Armed Services. We need to make new laws that protect our servicemen and women. We need to be allowed to sue when an injustice is done to us — yes, us because we are a unit. The military used to say and maybe still does, that if the military wanted you to have a spouse, they would have issued you one. Without the love of your family, it is hard to survive in a world of uncertainties. Knowing that there are loved ones taking care of their families at home, allows servicemen or women peace of mind. Loving them and supporting them gives these soldiers peace, knowing the families will receive them with open arms. These men and women have given their lives to save our lives. Yet they have no say in what happens in theirs.

Some friends have asked, "Aren't you afraid that they will try to kill you?"

I respond, "They already did when they took my husband from me. I have the armor of God on and He will protect me and only He will dictate when I die."

Thank you Jesus, amen!

The Armor of God is a metaphor in the Bible.

"Finally, be strong in the Lord and in his mighty power. Put on the full armor of God, so that you can take your stand against the devil's schemes. For our struggle is not against flesh and blood, but against the rulers, against the authorities, against the powers of this dark world and against the spiritual forces of evil in the heavenly realm. Therefore, put on the full armor of God, so that when the day of evil comes, you may be able to stand your ground, and after you have done everything, to stand. Stand firm then, with the belt of truth buckled around your waist, with the breastplate of righteousness in place, and with your feet fitted with the readiness that comes from the gospel of peace. In addition to all this, take up the shield of faith, with which you can extinguish all the flaming arrows of the evil one. Take the helmet of salvation and the sword of the spirit, which is the word of God. And pray in the Spirit on all occasions with all kinds of prayers and requests. With this in mind, be alert and always keep on praying for all the Lord's people." Ephesians 6:10-18 (New International Version)

Ronald Lee Treinen, would NEVER EVER betray his country! If asked, do I believe my husband was killed? The answer is yes!

"God Blesses those who work for peace, for they will be called the children of God."                    Matthew 5:9. ( New Living Translation)

MY HUSBAND, RON, IS EITHER LIVING HERE ON EARTH WITH A NEW LIFE AND IDENTITY OR IN THE LOVING ARMS OF THE SON OF GOD.

P.S. All people have secrets, I have a few too!!!!! ☺

# One last Kiss

I wanted one more kiss from you,
so I just ran outside.
But saw you drive away for school,
I'll let this time slide by.

I knew that you would come back home,
and we would kiss again.
So, I'll just wait until that time,
when you come home my friend.

The kiss would never ever happen,
'cause you did not come home.
I got the news of your demise
from a nurse on the phone.

We put you in a handsome box,
which they lowered in the ground,
and shoveled all the dirt on you.
To me, it was profound.

I only hope you understand,
to run and get that kiss.
Because a day may come for you,
that kiss, that you have missed.

Written by

Johanna M. Treinen

People say that you cannot get information from the Government, but that needs to change. If the government can sue the President of the United States, why can't we sue the military? We need some people who care and are not afraid to speak up. We need our government officials to fight for us and change some laws. Courageous people who want answers get noticed and the news picks up their stories. Then, it's all over the news and television. I'm not talking about frivolous things. I'm talking about killing our loved ones for unfounded fear and lying about what actually happened. Far too many unexplained deaths do not make sense. Stalking and trying to scare us to keep our mouths closed has to stop. Often people have loved ones die and no one knows what happened or where they went. We can either stand up for what is right and moral, as a tribute to our loved ones, or lie down and take it. I don't know about the rest of you, but I'm standing up, all the way up!

Blessings, Jo Treinen

# ABOUT THE AUTHOR

I am a sixty-nine-year-old woman, living in Virginia. I was born in Hawaii. My father was Puerto Rican and mother German. I am the fourth child of five children. I grew up in Chula Vista, California. I got married to a Marine when I was seventeen. No, I wasn't pregnant. I was in love and also wanted to get away from my father because he was strict and physically abusive. After nine years of the same treatment with my ex-husband, I filed for a divorce. I had three children from that union.

I married a sailor in 1982 and my life changed for the better. He had three girls and with my three, we raised six children together. With hard work, dedication and love, we made it through. Thank you, God! We traveled all over the United States. The Navy was good to us until the end of Ron's career. He died seven months after his retirement. I believe the military killed him. My story is about our lives and the after math of his unexpected death. I will never get that one last kiss. You will understand what I mean after reading my story.

Made in the USA
Monee, IL
09 August 2022

108d75a3-eca8-4ca6-bca6-64b3f2bfcb3dR02